TELL ME HOW IT WORKS

HOW DO THERMOMETERS WORK?

Clifton Park - Halfmoon Public Library
475 Moe Road
Clifton Park, NY 12065

Published in 2021 by The Rosen Publishing Group, Inc.
29 East 21st Street, New York, NY 10010

Copyright © 2021 by The Rosen Publishing Group, Inc.

All rights reserved. No part of this book may be reproduced in any form without permission in writing from the publisher, except by a reviewer.

First Edition

Editor: Siyavush Saidian
Book Design: Reann Nye

Photo Credits: Cover Patrick Daxenbichler/Shutterstock.com; Series Art (gears) goodwin_x/Shutterstock.com; Series Art (newspaper) Here/Shutterstock.com; p. 5 DeepDesertPhoto/RooM/Getty Images; p. 7 (top) https://commons.wikimedia.org/wiki/File:Justus_Sustermans_-_Portrait_of_Galileo_Galilei,_1636.jpg; p. 7 (bottom) flaviano fabrizi/Shutterstock.com; p. 8 videophoto/E+/Getty Images; p. 9 GinkoLac/Shutterstock.com; p. 11 Tomas Ragina/Shutterstock.com; p. 13 (top) Africa Studio/Shutterstock.com; p. 13 (bottom) Science & Society Picture Library/SSPL/Getty Images; p. 14 Eaum M/Shutterstock.com; p. 15 Nikita Rublev/Shutterstock.com; p. 17 eyedias/iStock/Getty Images Plus/Getty Images; p. 19 (top) adamkaz/iStock/Getty Images Plus/Getty Images; p. 19 (bottom) ChameleonsEye/Shutterstock.com; p. 21 Kriengsak tarasri/Shutterstock.com; p. 22 bixstock/Shutterstock.com.

Library of Congress Cataloging-in-Publication Data

Names: Mikoley, Kate, author.
Title: How do thermometers work? / Kate Mikoley.
Description: New York : PowerKids Press, [2021] | Series: Tell me how it works | Includes bibliographical references and index.
Identifiers: LCCN 2020001373 | ISBN 9781725318250 (paperback) | ISBN 9781725318274 (library binding) | ISBN 9781725318267 (6 pack)
Subjects: LCSH: Thermometers–Juvenile literature.
Classification: LCC QC271.4 .M55 2021 | DDC 681/.2-dc23
LC record available at https://lccn.loc.gov/2020001373

Manufactured in the United States of America

CPSIA Compliance Information: Batch #CSPK20. For Further Information contact Rosen Publishing, New York, New York at 1-800-237-9932.

CONTENTS

LET'S TELL TEMPERATURE! 4
THE FIRST THERMOMETER? 6
LOTS OF LIQUID 8
ONWARD AND UPWARD 10
FAHRENHEIT AND CELSIUS 12
LIMITS OF LIQUIDS 14
DIAL THERMOMETERS 16
ELECTRONIC THERMOMETERS 18
THERMOCOUPLES 20
WHAT'S NEXT? 22
GLOSSARY 23
INDEX 24
WEBSITES 24

LET'S TELL TEMPERATURE!

Temperature is a measurement that tells us how hot or cold something is. To find a temperature, we use a tool called a thermometer. Different kinds of thermometers can find the temperatures of different things. Thermometers can measure the temperature of air, water, food, and even your body!

Units used to measure temperature are called degrees. Most Americans use a temperature scale called Fahrenheit. In other parts of the world, people measure temperature using the Celsius scale. Even in the United States, scientists often use Celsius.

Water freezes at 32° Fahrenheit (0°Celsius). It boils at 212°F (100°C).

THE FIRST THERMOMETER?

In the late 1500s, a tool was invented that some people think is the first thermometer. A man from Italy named Galileo Galilei studied math and science. Around 1592, he built a tool made of glass. It was filled with liquid that moved to show a change in temperature.

Galileo's invention is actually called a thermoscope. While many people think of it as the first thermometer, it didn't have a scale to measure exact temperatures. As such, it wasn't a true thermometer.

TECH TALK

While Galileo's invention wasn't exactly a thermometer, the basics of it were important. Other scientists would work off of Galileo's ideas to come up with thermometers like the ones used today.

GALILEO GALILEI

A few years after Galileo invented his thermoscope, other scientists were able to add numbers to it for a scale. These were the first true thermometers.

LOTS OF LIQUID

Soon, many scientists were coming up with their own thermometers. For the most part, they all worked in the same basic way as Galileo's. A liquid expands, or gets bigger, as it gets hotter. As it cools, it contracts, or gets smaller. By putting liquid inside a **container** and measuring how much it changes, temperature can be found.

MERCURY

TECH TALK

Basic thermometers once often contained, or held, a silver metal called mercury. At most temperatures, mercury is a liquid. It expands at an even rate, making it useful for thermometers.

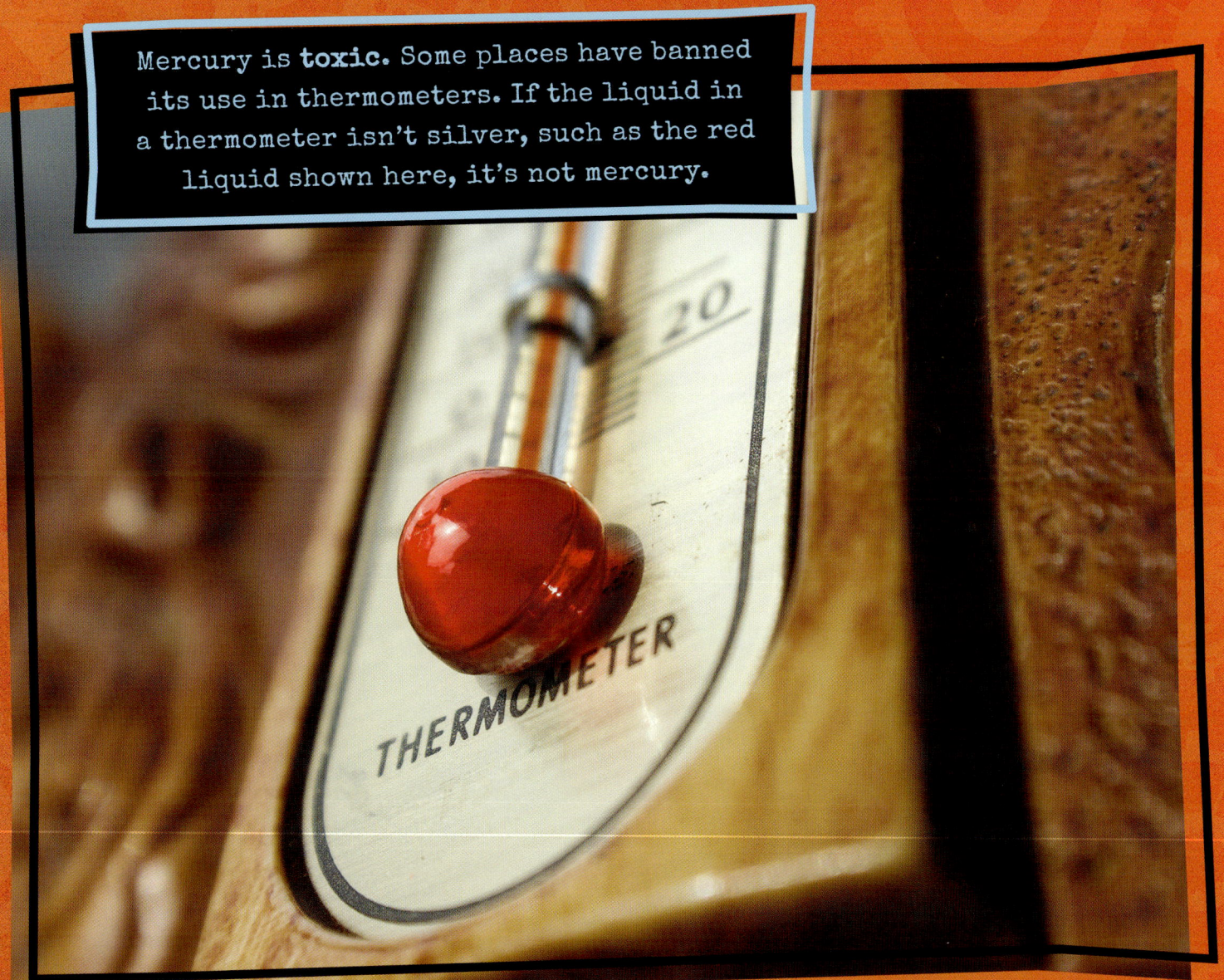

Mercury is **toxic**. Some places have banned its use in thermometers. If the liquid in a thermometer isn't silver, such as the red liquid shown here, it's not mercury.

By the 1700s, scientists had tried many different liquids in thermometers. They had also tried a lot of different temperature scales.

ONWARD AND UPWARD

Today, one of the simplest kinds of thermometers is a glass tube filled with liquid. This is called a liquid thermometer. There's a rounded part, called the bulb, at the bottom of the tube.

The thermometer's tube is small and sealed. There's only a certain amount of room for the liquid to move. As the temperature gets hotter or colder, the liquid inside the tube must expand or contract. Since the tube is sealed, the only direction expanding liquid can go is upward.

> The liquid in the thermometer is held inside a tube called the stem. Scales along the side often show the temperature in both Celsius and Fahrenheit.

PARTS OF A LIQUID THERMOMETER

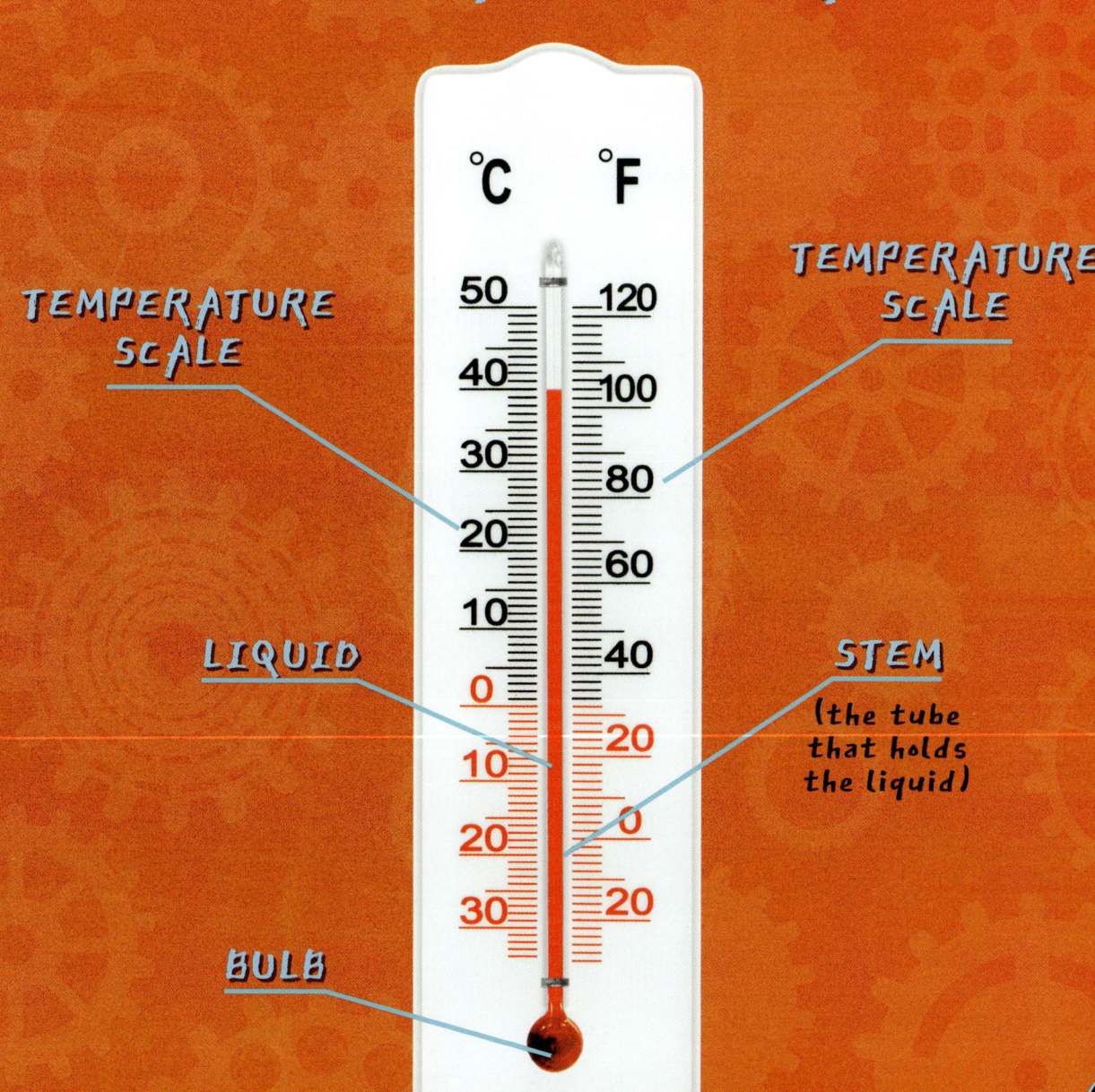

FAHRENHEIT AND CELSIUS

In the early 1700s, a scientist named Daniel Fahrenheit made a mercury thermometer that **accurately** measured temperature based on his own scale. The scale, now known as the Fahrenheit temperature scale, was based on a measurement of 32 degrees for water's freezing point and 212 degrees for its boiling point. The space between the two temperatures was **divided** into 180 equal parts, or degrees.

In 1742, another scientist, Anders Celsius, came up with a scale made up of 100 degrees. It's called a centigrade scale.

TECH TALK

Celsius first used 0°C to mark the boiling point of water and 100°C to mark snow's melting point. Today, this is flipped—0°C is water's freezing point and 100°C is its boiling point.

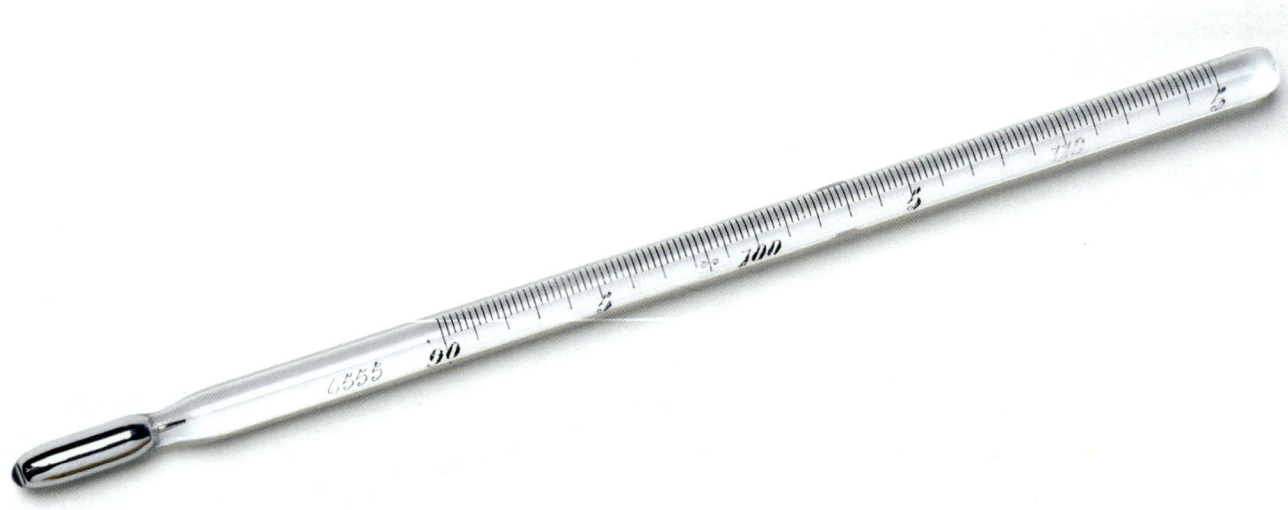

In addition to the mercury thermometer, Fahrenheit also created a thermometer using alcohol.

LIMITS OF LIQUIDS

The type of liquid used in a thermometer can limit how hot or cold of a temperature it can measure. Mercury, for example, turns to a solid at -37.9 °F (-38.8 °C). This means the liquid would freeze and can't be used to measure temperatures below this point.

TECH TALK

Liquids aren't the only **substances** used in thermometers. Gas thermometers work mostly the same way as liquid thermometers. They work especially well for cold temperatures.

To find accurate temperatures, scientists have to consider how fast the substance being used expands or contracts.

Alcohols commonly used in thermometers boil at about 172°F (78°C). This means that once the object being measured gets hotter than this, alcohol thermometers can't measure the temperature.

DIAL THERMOMETERS

Not all thermometers work the way liquid thermometers do. You've probably seen a **dial** thermometer. These are often used to measure outdoor and food temperatures. They look a bit like a clock, with a hand that points to a place on a circle to show the temperature.

Dial thermometers work by using two metals that expand at different rates as they're heated. These metals are made into a strip that expands and bends as it heats, pushing the hand further around the dial.

Outdoor dial thermometers often work by having the metal strips shaped like **coils**. This allows the thermometer to sense even very small changes in temperature.

ELECTRONIC THERMOMETERS

Another kind of thermometer is the electronic, or digital, thermometer. These have many uses, including taking your body temperature. They commonly have a small sensor made of metal. The **resistance** in this metal changes as the temperature changes. A **microchip** inside the thermometer reads the measurement of resistance and changes it into whichever temperature scale the thermometer uses.

These thermometers can give you a temperature reading very quickly. The temperature pops up on a screen, making it very simple to use.

TECH TALK

Digital thermometers are often used to make sure your body temperature is in the healthy range. Normal body temperature is about 98.6°F (37°C). If you're sick, your temperature may be hotter.

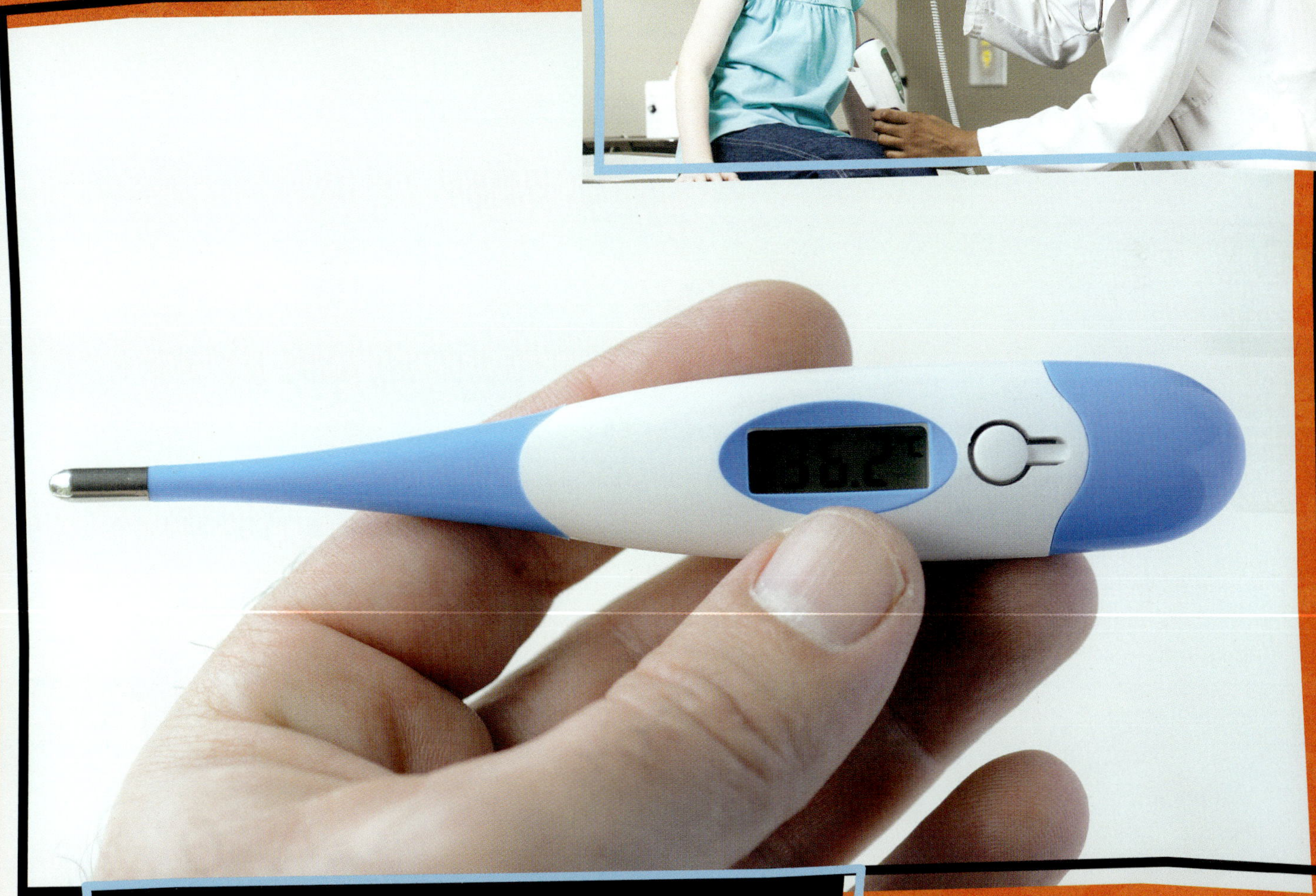

The doctor may use a digital thermometer to check your temperature when you go for a checkup.

THERMOCOUPLES

Some objects are too hot for regular thermometers to measure. Likewise, some things are too cold. To find these temperatures, scientists use a special thermometer called a thermocouple.

A thermocouple has two wires that join together. Each is made of a different metal. One end of the wires is connected to a tool that measures **voltage**. The other is placed where the temperature is being measured. When the different metals are heated or cooled, they produce a voltage, which is used to find the temperature.

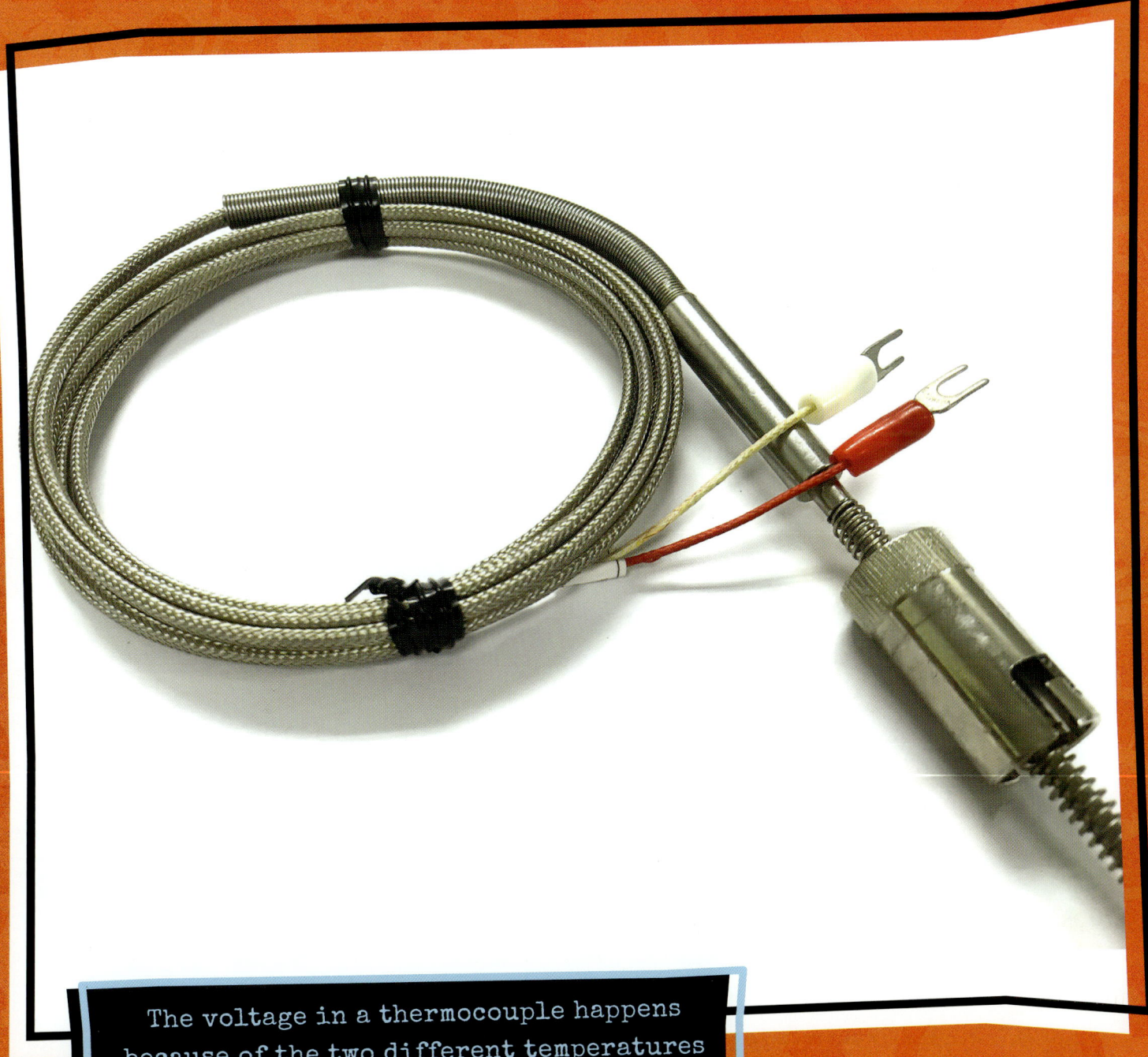

The voltage in a thermocouple happens because of the two different temperatures at each end of the wires.

WHAT'S NEXT?

Today, we have many types of thermometers. Some are very advanced, like the thermocouple. Cryometers are another kind of thermometer that can measure very low temperatures. They're mostly used to take temperatures in space. Pyrometers can measure very hot temperatures. They're often used to measure the temperatures of very hot metals.

Thermometers have come a long way since the ones developed right after Galileo's thermoscope. As **technology** continues to advance, so will the kinds of thermometers we use. Maybe someday you'll invent the newest thermometer!

GLOSSARY

accurate: Free from mistakes.

coil: A series of loops.

container: An object used to hold something.

dial: A round part of a tool with numbers or marks to show some measurement usually by means of a pointer.

divide: To break up.

microchip: A group of tiny electronic paths that work together on a very small piece of material.

resistance: The opposition a substance has to the passage through it of an electric current.

substance: A certain kind of matter.

technology: Tools, machines, or ways to do things that use the latest discoveries to fix problems or meet needs.

toxic: Harmful to human health.

voltage: A measurement of electrical energy.

INDEX

A
alcohol, 13, 15

C
Celsius, Anders, 12
cryometer, 22

D
dial thermometer, 16, 17

E
electronic (digital) thermometer, 18

F
Fahrenheit, Daniel, 12, 13

G
Galilei, Galileo, 6, 7, 8, 22

L
liquid, 6, 8, 9, 10, 14, 16

M
mercury, 8, 9, 12, 13

P
pyrometer, 22

S
scale, 4, 6, 7, 8, 9, 12, 18

T
thermocouple, 20, 21, 22
thermoscope, 6, 7, 22

Due to the changing nature of Internet links, PowerKids Press has developed an online list of websites related to the subject of this book. This site is updated regularly. Please use this link to access the list: www.powerkidslinks.com/tmhiw/thermometers

Clifton Park-Halfmoon Library

0000605518026